STEAM LOCOMOTIVES OF THE B & O RAILROAD

PHOTO ARCHIVE

Kim D. Tschudy

Iconografix
Photo Archive Series

Iconografix
PO Box 446
Hudson, Wisconsin 54016 USA

Library of Congress Card Number: 2003103522

ISBN 1-58388-095-X

03 04 05 06 07 08 09 5 4 3 2 1

Printed in China

Cover and book design by Dan Perry

Copyediting by Jane Mausser

COVER PHOTO: The Baltimore & Ohio 4-6-0 ten-wheeler locomotive on display at the 1939 New York World's Fair. The depression was finally drawing to an end bringing renewed hope to America. That hope was short-lived. Within two short years the United States would be involved in WWII. The B & O advertised that passengers could visit not only the New York World's Fair but also Washington, "America's great shrine of patriotism," as well as a visit to Philadelphia, "America's cradle of liberty."

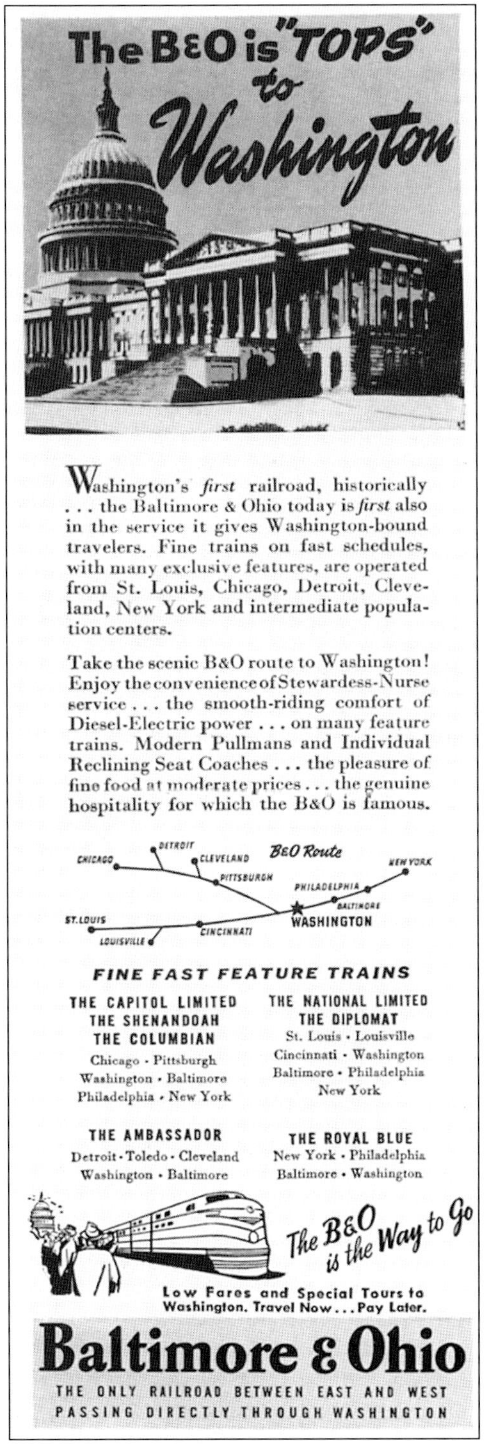

Table of Contents

Dedication

I dedicate this book on the Baltimore & Ohio Railroad to the memory of Gilpin A. Morrison. Morrison was working in the B & O yard at Connellsville, Pennsylvania, on March 5, 1890, when a train struck and killed him.

It was through his great-great granddaughter, Karen Morrison, that I learned of Morrison's life and untimely death while employed by the Baltimore & Ohio. I also dedicate this book to the memories of all the railroad workers across the country who gave their lives and limbs for the building of America. Since 1827, the railroad workers have been the unsung heroes of America.

Thank you to those friends near and far who have encouraged me, put up with me, and tolerated me during the writing of this book. You know who you are. Thanks for having the confidence in me that kept me writing and searching for interesting information about the railroad that began railroading in America.

Acknowledgments

All photographs in this book are from the Mike Gruber collection. Thank you, Mike, for helping to preserve the history of the B & O with these wonderful, detailed photographs.

Introduction

Baltimore Harbor, November 12, 1993.

Standing on the main deck of the U.S. Navy hospital ship U.S.S. *Sanctuary* I can see both the flag flying over Fort McHenry and the Baltimore & Ohio Railroad's Locust Point yards.

Nearby are several women walking the gangplank to reach the main deck. They are obviously reflecting on their pasts. For many of these women the *Sanctuary* represented their first nursing job after graduation from nursing schools all across America. Now middle-aged, these women were but 20 years old when they first boarded the *Sanctuary* for its weekly roundtrips from the Delta to the DMZ in the coastal waters of Vietnam. Yesterday was spent at the dedication of the Women's Vietnam Veterans Memorial in Washington.

Watching the flag over Fort McHenry stirs many emotions from deep within. My vantage point gives me a chance to reflect and contemplate; this flag symbolizes a new nation, freedom and the ideas that live today, and hope for future generations...

Throughout the nineteenth century, ships sailing into Baltimore harbor brought immigrants from all corners of the globe. Entering the harbor was significant for them. It was the end of one chapter of their lives and the beginning of a new life in the land of hope and promise.

Once in Baltimore, countless numbers of immigrants found a railroad that has since stood the test of time. This railroad gave them hope and helped them fulfill their dreams. While time has faded the mark of the B & O, many families that trace their history back to the old country, including mine, learn that the B & O brought their families to America's heartland.

The history of the Baltimore & Ohio Railroad is the history of America. In 1828 at age 91, Charles Carrol of Carrolton, Maryland, laid the first stone to begin the construction of the railroad. Carrol was the last living signer of the Declaration of Independence.

By 1835, America's railroad had opened its line to Washington. The first trip was made the same day construction of the Thomas Viaduct was completed. Within a short time four locomotives, the *George Washington*, *John Adams*, *Thomas Jefferson*, and the *James Madison* were in regular service between Baltimore and Washington. Three years later, the railroad began carrying the mail between Baltimore and Washington.

In 1843, the first telegraph wire was laid along the railroad right-of-way in preparation for an experiment the following year. On May 24, 1844, communications changed forever when Samuel F. B. Morse sent the first telegram. It was sent from the basement of the U.S. Supreme Court building in Washington to the Mount Clare Depot, 40 miles away.

In 1859, troops were quickly dispatched via the B & O to Harper's Ferry after John Brown's raid on the Union Arsenal. Two years later the railroad became a major force in the Civil War. During that conflict, the B & O was called upon repeatedly to move equipment, horses, and men to the battlefield. Within weeks of the conclusion of the Civil War, the railroad was once more called to active duty to bring the body of the assassinated president, Abraham Lincoln, back home to Springfield, Illinois.

It was in wartime that the true personality and importance of the B & O Railroad emerged.

Confederate General Stonewall Jackson realized early in the war that the B & O line to Wheeling, West Virginia was a most important Union supply line. With that realization and the potential the line held for a quick re-supply of Union troops Jackson embarked on a campaign to disrupt the line that led to the mouths of three important valleys. The campaign for the most part was a failure.

Numerous towns in West Virginia changed hands over the course of the war because of their important railroad links. Romney, West Virginia was one such town that changed hands many times during the Civil War.

In every war since the Civil War, the Baltimore & Ohio has stood tall when called upon by America.

Hardly a child today hasn't heard of the B & O Railroad, even if only from the game "Monopoly." However, as children grow and begin their studies of American history they will learn about the race between the *Tom Thumb* and a horse-drawn carriage.

The B & O has laid tracks into history, stirring memories and nostalgia for many Americans today. It is a testament to the railroad's importance that the Baltimore & Ohio can still warm something deep within the soul.

Tom Thumb (right) was the first locomotive built for the B & O by Peter Cooper. Phineas Davis, a watchmaker, built the *Atlantic* (left) in York, Pennsylvania, in 1832. This pair of locomotives was photographed on July 28, 1940, while on display at the New York World's Fair.

This is the *Atlantic* and two of the earliest Baltimore & Ohio passenger cars. Imagine what a ride it must have been to be sitting on the top bench seats. The passenger cars look like a cross between a stagecoach and an early sailing ship. It was photographed at the Chicago World's Fair in 1948–1949.

No. 97, a Class C-16 0-6-0T six-wheel tank switcher, was built by Baldwin Locomotive Works in 1912. It was photographed on October 6, 1948, in Baltimore, Maryland. The locomotive was renumbered No. 897; in 1950, it was scrapped at the Mount Clare Shops.

The Rhode Island Locomotive Works built No. 320, a Class D-22 0-6-0T six-wheel tank switcher, in 1892. Photographed in Ivorydale, Ohio, on April 28, 1947, this locomotive is the ex-No. 57. It was renumbered several times—first to the number shown on the photo, later to No. 1120 in 1950.

No. 380, a Class D-30 0-6-0 six-wheel switcher, was photographed on October 19, 1947 in Lima, Ohio. Freight from Fredrick to Baltimore on January 28, 1832, consisted of 52 carloads including 852 barrels of flour, 20 tons of granite, 12 barrels of glue, 4 tons of bark, and 10 tons of wood.

No. 384, a Class D-30 0-6-0 six-wheel switcher, was photographed on March 14, 1939, in Glenwood, Pennsylvania. In 1832, just three months after the railroad line opened to Frederick, Maryland, it carried 32,670 barrels of flour from Frederick.

In 1919, the Baldwin Locomotive Works built the Class D-30 0-6-0 six-wheel switcher No. 1151. This locomotive was photographed in Dover, Ohio, in 1952. Philip E. Thomas, the first president of the B & O Railroad, was appointed commissioner of the Chesapeake & Ohio Canal Company to represent the interests of the state of Maryland.

0-8-0s

On February 12, 1938, C. Dengler photographed No. 613, a Class L-2 0-8-0 eight-wheel switcher, in Glenwood, Pennsylvania. Note the absence of snow in the photograph. Within one year of his appointment as commissioner of the Chesapeake & Ohio Canal, Philip Thomas resigned from that position, determined that the proposed canal was a futile effort and that another means of providing commerce to Baltimore could be found.

No. 624, a Class D-30 0-8-0 eight-wheel switcher, is shown in Glenwood, Pennsylvania, on March 4, 1939. Originally built as B & O No. 3542, Class E-27, it was rebuilt in July 1925 at the Mount Clare Shops. Note the engineer posing for the unknown photographer.

No. 662, a Class L-2b 0-8-0 eight-wheel switcher, was photographed on May 12, 1947, at Fairmont, West Virginia. Originally built as B & O No. 2579, the Class E-27 2-8-0 locomotive was rebuilt in August 1926 at the Mount Clare Shops. Many Irish immigrants populated Fairmont in the 1850s and worked in the coalmines and for the B & O Railroad.

On April 28, 1947, No. 692, a Class L-2a 0-8-0 eight-wheel switcher, was photographed at Ivorydale, Ohio. Originally built as Class E-27b 2-8-0 locomotive No. 2791, it was rebuilt in April 1929 at the Mount Clare Shops. Incidentally, the world famous Ivory soap is made in Ivorydale. The B & O hauled freight into the Proctor & Gamble factory that manufactured Ivory soap.

On September 29, 1957, Paul Stringham photographed No. 700, a Class L-2 0-8-0 eight-wheel switcher, at Riverdale, Illinois. Note the B&OCT lettering on the front of the locomotive. The B&OCT was the Baltimore & Ohio Chicago Terminal Railroad.

No. 712, a Class L-2c 0-8-0 eight-wheel switcher, was photographed on May 8, 1950, in New Castle, Pennsylvania. This locomotive was originally built as a Class E-27 2-8-0, No. 2545; it was rebuilt in 1941 at the Mount Clare Shops.

No. 1023, a Class L-1a 0-8-0 eight-wheel switcher, was photographed on April 24, 1947, at Grafton, West Virginia. Originally built as a Class E-24 2-8-0, locomotive No. 2211 was rebuilt in the Mount Clare Shops in March 1924. Today the Mount Clare Shops serves as the Baltimore & Ohio museum.

Charles T. Felstead photographed No. 1653, a Class L-2b 0-8-0 eight-wheel switcher, on September 14, 1956, at Willard, Ohio. The town of Willard, Ohio was originally named Chicago, Ohio. It was changed to Chicago Junction but this still created problems for the U.S. postal service. The town asked permission to change the name to Willard to honor Daniel Willard, president of the Baltimore & Ohio Railroad. The new name became official in September 1917.

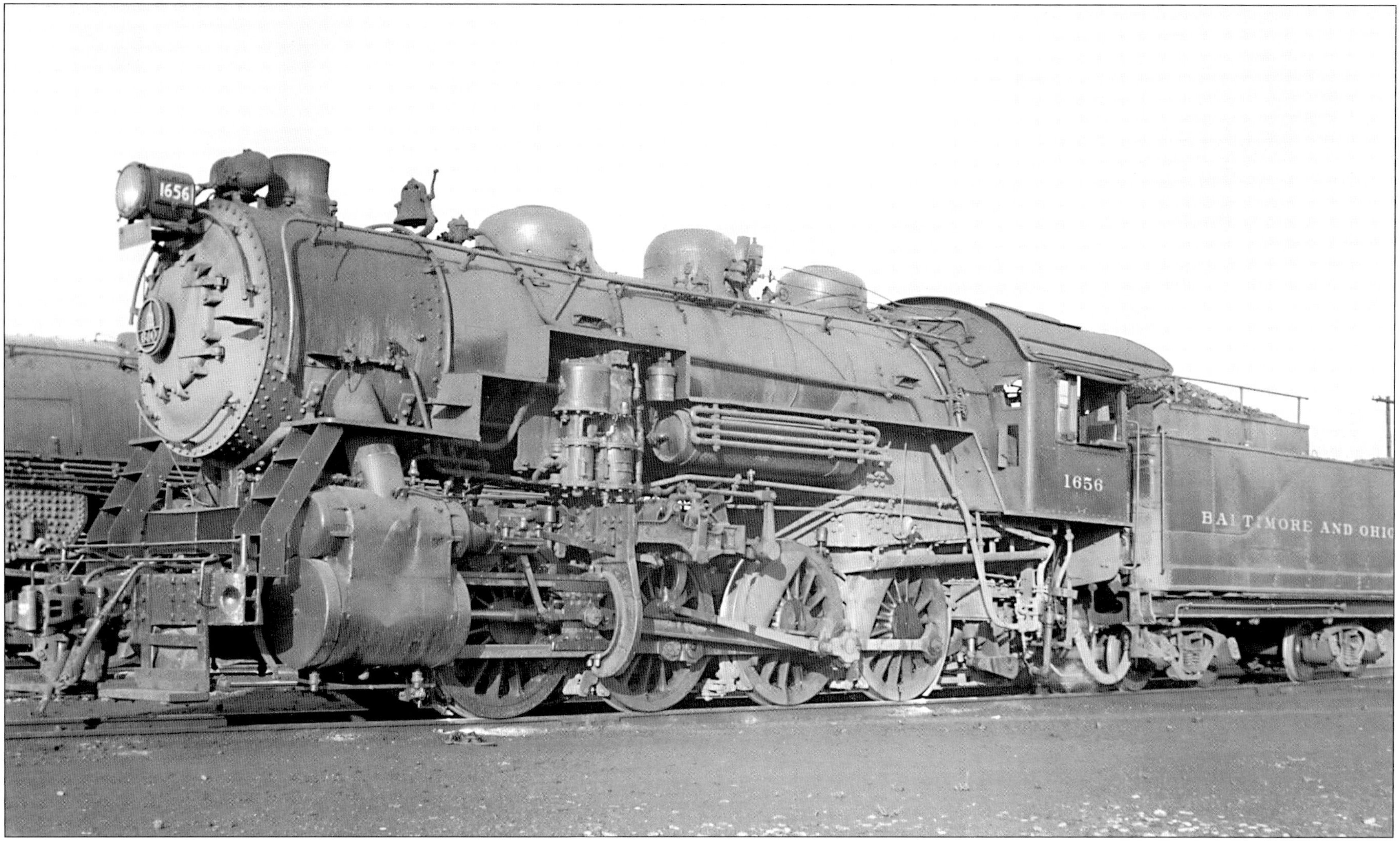

Charles T. Felstead also photographed No. 1656, a Class L-2 0-8-0 eight-wheel switcher, on July 22, 1956, at Willard, Ohio. Willard was founded primarily to serve the needs of the Baltimore & Ohio Railroad, which provided jobs for Willard area residents.

No. 1074, a Class L-1a 0-8-0 eight-wheel switcher, was photographed on May 7, 1947, at Glenwood, Pennsylvania. This locomotive was originally built as B & O locomotive No. 2246, a Class E-24 2-8-0, which was rebuilt in March 1928 at the Mount Clare Shops.

0-8-8-0s

No. 7044, a Class LL-1 0-8-8-0 articulated, was photographed on May 12, 1949, at Brunswick, Maryland. It was built by Alco Schenectady in 1913 as a Class 0-1 No. 2426 and later renumbered to its current number. This locomotive was scrapped in 1950. Brunswick, Maryland is home to the Brunswick Railroad Museum that features the Baltimore & Ohio.

0-10-0s

On April 23, 1947, No. 950, a Class U 0-10-0 ten-wheel switcher, which saw use in a number of B & O yards, was photographed in Cumberland, Maryland. Originally B & O No. 6009 Class S 2-10-2, it was rebuilt in June 1925 at the Mount Clare Shops with the running gear from locomotive No. 6009 and a new boiler. It was renumbered as No. 1950 in 1951.

Charles T. Felstead photographed No. 951, a 0-10-0 ten-wheel switcher, on September 30, 1949, at New Castle, Pennsylvania. In November 1842, the opening of the line to Cumberland was greeted with great joy as it meant the B & O had now positioned itself for its move into Ohio.

2-6-0s

No. 600, a 2-6-0, Mogul-type locomotive, is shown here at the New York World's Fair in 1939. In 1848, a number of letters signed by "A Large Stockholder," critical of the B & O began appearing in the Baltimore *Patriot*. In time, the identity of the writer became known, and he, Thomas Swann, was elected to the board of directors of the railroad in October 1848.

No. 904, a 2-6-0 Mogul-type locomotive, was photographed in Chicago, Illinois. In 1964, the Baltimore & Ohio Chicago Terminal (B&OCT) Railroad had 81 route miles, 365 track miles, and 38 locomotives.

This 2-6-0 Mogul-type B&OCT locomotive, No. 905, was photographed in Chicago. The B&OCT was a separate company fully owned by the B & O. The Pere Marquette, B & O, Soo Line, and Chicago & Great Western railroads' passenger trains used B&OCT tracks to get to downtown Chicago.

No. 2437, a 2-6-0 Mogul-type locomotive, photographed on November 27, 1937, in Baltimore, Maryland. Built by the Baldwin Locomotive for the Chicago & Alton Railroad as No. 2437, a Class K-27 locomotive, it later became a B & O locomotive of the same number. It was scrapped at Ivorydale, Ohio, in May 1947.

2-6-6-2s

On May 11, 1947, No. 7516, a 2-6-6-2 Class KK-4b articulated locomotive, was photographed at Du Bois, Pennsylvania. It was built in 1918 by the Alco-Brooks Locomotive Works as No. 58002 for BR & P Railroad, which renumbered it as No. 716 Class LL. It was scrapped in 1951.

No. 7520, a 2-6-6-2 Class KK-b4 articulated locomotive, was photographed on June 8, 1947, at Punxsutawney, Pennsylvania. Because of the size of the articulated locomotives, a number of B & O shops had a difficult time servicing them.

H. Newlon photographed No. 7529, a 2-6-6-2 Class KK-4b articulated locomotive. When William G. Harrison, the fifth president of the B & O Railroad, took office in 1853, he was a businessman and not a railroader. He didn't assume the presidency for the money; his annual salary was $3,000 per year.

2-6-6-4s

The articulated locomotive No. 7700, 2-6-6-4 Class KB-1, was photographed on May 12, 1949, in Brunswick, Maryland. It was built by Baldwin Locomotive Works as No. 61829 in April 1935. This locomotive was originally owned by Seaboard Air Line Railroad and operated as No. 2500, a Class R-1 locomotive.

No. 7703, a 2-6-6-4 Class KB-1 articulated locomotive, was photographed on May 11, 1949, at Keyser, West Virginia. The arrival of the B & O took Keyser from a quiet rural town to a thriving industrial center.

No. 7703 photographed on May 12, 1949, at Brunswick, Maryland. It was built by Baldwin in April 1935 as No. 61832 and originally owned by Seaboard Air Line Railroad. Seaboard operated it as No. 2503, a Class R-1 locomotive.

No. 7709 is a 2-6-6-4 Class KB-1a articulated locomotive. It was photographed on October 7, 1948, at Cumberland, Maryland. The Baldwin Locomotive Works built it in August 1937 and it was originally owned by Seaboard Air Line Railroad and operated as No. 2509 Class R-1.

2-8-0s

No. 215, a 2-8-0 consolidation-type locomotive, was photographed on June 22, 1951, at Columbus, Ohio. By 1857, construction of the Central Ohio Railroad was complete. The Central Ohio linked the cities of Bellaire, Zanesville, Newark, and Columbus. These connections would provide the B & O Railroad with much freight and many passengers.

Shown on August 6, 1939, at Glenwood, Pennsylvania, is No. 1278, a 2-8-0 Class E-14 consolidation locomotive. The missing headlamp on this locomotive was probably in storage at the time the photo was taken.

No. 2291 2-8-0 Class E-24a is a consolidation-type locomotive. It was photographed on August 12, 1946, at Glenwood, Pennsylvania. Judging from the 4 to 5 story building in the background this locomotive must have been working in the industrial area of Glenwood.

Robert J. Foster photographed No. 2563, a 2-8-0 consolidation-type locomotive, on April 4, 1950, at East St. Louis, Illinois. Locomotive No. 2563, Class E-27, was built by Alco Schenectady Locomotive Works as No. 38296 in November 1906 and scrapped in May 1954.

Shown on October 16, 1938, at Pittsburgh is No. 2769, a 2-8-0 Class E-27ca consolidation-type locomotive. The Richmond Locomotive Works built it as Richmond No. 47147 in March 1910. It was scrapped in 1950. The line from Port Perry to Pittsburgh was completed and opened for traffic in October 1861.

Photographed on June 14, 1952, at Springfield, Illinois, is No. 2796, a 2-8-0 Class E-27a consolidation-type locomotive. On the 100th anniversary of the B & O Railroad, the Springfield Division operated on a 228-mile line from Beardstown to Shawneetown, Illinois.

No. 2807, a 2-8-0 Class E-27ca consolidation-type locomotive, photographed by Paul Stringham on September 30, 1952, at Decatur, Illinois. Still in Decatur is a pair of old B & O coach or baggage cars in their original blue and gray colors. They may be in use as a yard office.

No. 2818, a 2-8-0 consolidation-type locomotive, was photographed on May 26, 1945, at New Castle, Pennsylvania. No doubt this locomotive pulled thousands of tons of freight to help in the Allied war effort. Note the well-kept depot in the background.

No. 2855, a 2-8-0 Class 27ca consolidation-type locomotive, was photographed on October 16, 1949, at Decatur, Illinois. The Decatur depot was used by the Illinois Central, C & IM, B & O, and the Springfield & Southwestern, a shortline that went broke before World War II.

On October 11, 1953, R. J. Foster photographed No. 2908, a 2-8-0 Class E-27ca consolidation-type locomotive, at Flora, Illinois. Flora was platted for establishment of a railroad depot connecting Cincinnati and East St. Louis. It was named after the daughter of the railroad surveyor.

W. Krambeck photographed No. 2917, a 2-8-0 Class 31b consolidation-type locomotive, on March 5, 1950, at East St. Louis, Illinois. The 335-mile line from Cincinnati to East St. Louis was built by H. C. Seymour at a cost of $9 million, which included all depots, station houses, turn tables, and machine shops.

Photographed on July 18, 1947, in Philadelphia, Pennsylvania, is No. 2929, a 2-8-0 Class E-31aa consolidation-type locomotive. The CH& D Railroad formerly owned this locomotive. The tender is an unusual design: short coal bunker and round water tank. Note the contemplative engineer looking out the window.

No. 3105, a 2-8-0 consolidation-type locomotive, was photographed on May 26, 1950, at New Castle Junction, Pennsylvania. William G. Harrison, fifth president of the B & O, served from 1853 to 1855.

Locomotive No. 4000, Class Q-1 was built in January 1911 by the Baldwin Locomotive Works as Baldwin No. 35911, rebuilt to a Class Q-1aa locomotive in 1927 and retired in March 1955.

No. 4275, a 2-8-2 Class Q-1c mikado-type locomotive, was photographed on April 21, 1950 at Riverdale, Illinois. This locomotive is also bearing a Baltimore & Ohio Chicago Terminal placard.

Photographed on August 19, 1956, at Clarksburg, West Virginia, is No. 4443, a 2-8-2 mikado-type locomotive. B & O President Leonor F. Loree purchased the Clarksburg Coal Company when it was learned that George Gould, acting through the Wabash Railroad, was attempting to purchase the Fairmont Coal Company. This sale would have meant additional competition for the B & O.

On April 4, 1939, No. 4457, a 2-8-2 mikado-type locomotive, was photographed at Chicago, Illinois. For many years, the B & O used the Illinois Central Passenger Depot located at Randolph Street in the Loop. The B & O built a 600-foot-long freighthouse nearby to accommodate the freight the B & O was moving into Chicago from points east.

No. 4469 is a 2-8-2 mikado-type locomotive. It was photographed on August 18, 1956, at Connellsville, Pennsylvania. Baltimore & Ohio locomotive No. 79 killed switchman Gilpin A. Morrison at Connellsville, on March 5, 1890. The Junior Mechanics and the Redmen, both of which Morrison was a member of, conducted funeral rites.

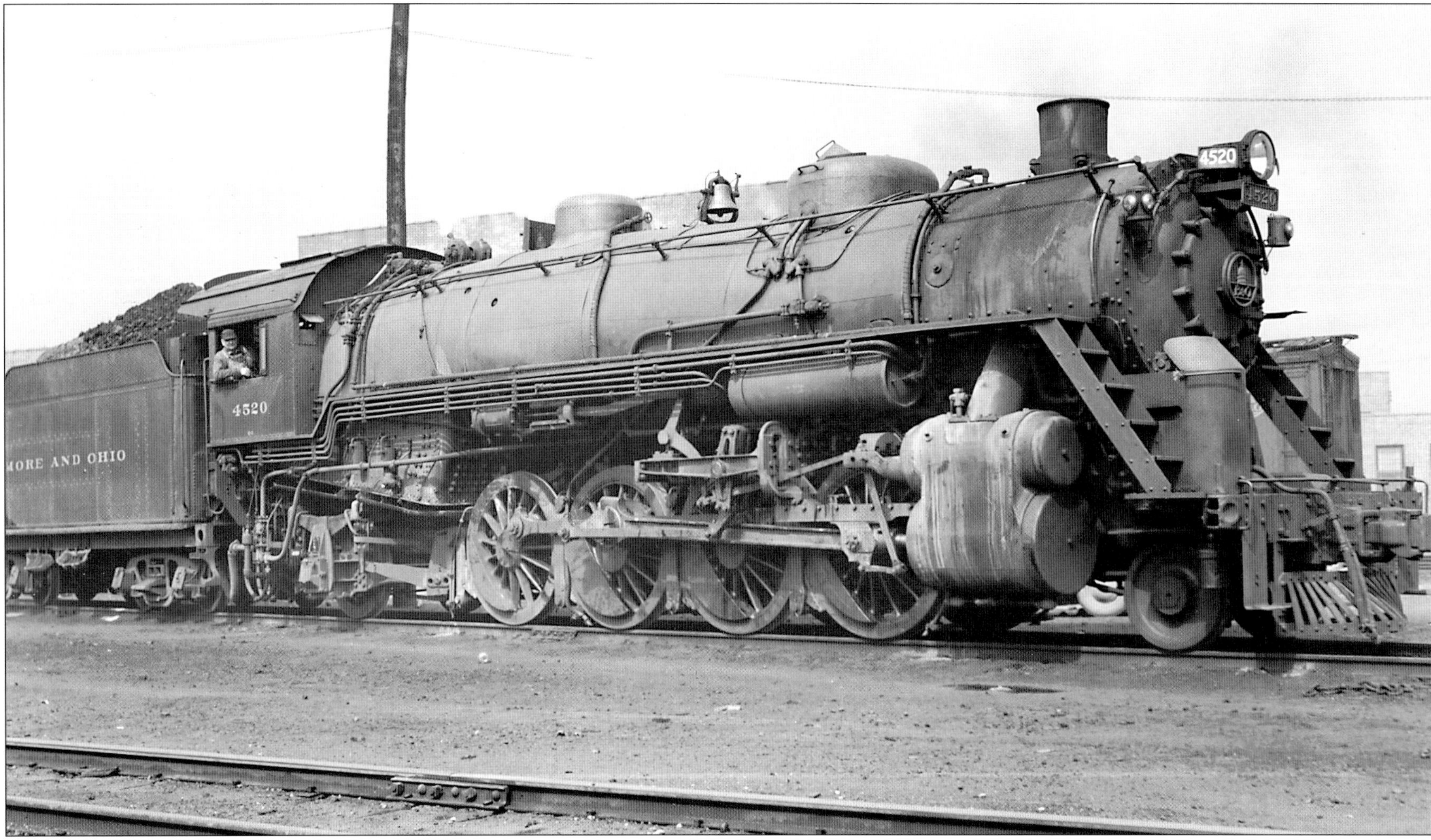

On March 20, 1948, R. J. Foster photographed No. 4520, a 2-8-2 Class Q-3 mikado-type locomotive, at East St. Louis, Illinois. This locomotive was renumbered to No. 320 in 1957. In 1939, Arthur Gates went to work as a freighthouse worker, eventually becoming a switchman at $7.32 per day. In 1943, Gates became the yard foreman at Cone Yard in East St. Louis.

R. J. Foster also shot No. 4577, a Class Q-3, 2-8-2 mikado-type locomotive, this time on February 18, 1948. Originally built as B & O No. 377 by Baldwin Locomotive Works as Baldwin No. 51144 in January 1919, it was retired in December 1957.

No. 4618, Class Q-4b locomotive, was photographed on August 19, 1956 at Grafton, West Virginia. This locomotive was built by Baldwin Locomotive Works in March 1923 and retired in November 1956.

H. Newlon photographed No. 4739 on August 1, 1939, a 2-8-2 Class Q-10b mikado-type locomotive built by Alco-Brooks as Alco No. 57200 in February 1917. It was retired in 1953, then sold for scrap the same year. Locomotives numbered 4707-4746 were all ex-BR & P Railroad locomotives.

No. 4877 is a 2-8-2, Class Q-7f mikado-type locomotive built by Baldwin Locomotive Works as Baldwin No. 44557 in November 1916 and sold for scrap in February 1953.

2-8-8-0s

On August 8, 1947, No. 7106, a 2-8-8-0 articulated locomotive, was photographed at M & K Junction, West Virginia. This Class EL-1 locomotive was built by Baldwin Locomotive Works as Baldwin No. 40371 in March 1916 and converted to a Class EL-1a locomotive between 1927-1932. It was sold for scrap in March 1951.

Photographed at Connellsville, Pennsylvania, on August 10, 1947, is No. 7122, a 2-8-8-0 Class EL-3. This Class EL-3 locomotive was built by Baldwin Locomotive Works as Baldwin No. 47282 in December 1917 and was scrapped in June 1956.

No. 7134, a 2-8-8-0 Class EL-3a articulated locomotive, was photographed on May 7, 1947, at Glenwood, Pennsylvania. This Class EL-3 locomotive was built by Baldwin Locomotive Works as Baldwin No. 48278 in March 1918 and rebuilt to a Class EL-3a between 1927-1932, then sold for scrap in June 1952.

No. 7139 is a 2-8-8-0 Class EL-3a articulated-type locomotive. This locomotive was built as a Class EL-3 by Baldwin Locomotive Works as Baldwin No. 48940 in June 1918 and rebuilt to a Class EL-3a locomotive between 1927-1932, then sold for scrap in June 1952.

On July 3, 1947, No. 7145, a 2-8-8-0 Class EL-5d articulated locomotive, was photographed at Glenwood, Pennsylvania. This locomotive was built as a Class EL-5 locomotive by Baldwin Locomotive Works as Baldwin No. 52371 in September 1919. It was rebuilt to a Class EL-5a between 1925-1930 and retired in 1954.

The Baldwin Locomotive Works built No. 7156, a 2-8-8-0 Class EL-5a, in 1919. It was photographed on May 10, 1947, at Glenwood, Pennsylvania. Railroaders and mill workers populated the towns of Glenwood and Hazelwood; typically, the railroad workers in Glenwood and the mill workers lived in Hazelwood.

In 1916, Alco Schenectady built No. 7212, a 2-8-8-0 Class EL-2 articulated. It was photographed at M & K Junction, West Virginia, on April 24, 1947. This locomotive was built by Baldwin Locomotive Works as Baldwin No. 55676 in April 1916 and sold for scrap in August 1951.

2-8-8-4s

No. 7606, a 2-8-8-4 Class EM-1 mallet-type locomotive, was photographed on June 4, 1949 at Keyser, West Virginia. The town of New Creek Station changed its name to just New Creek. In 1874, New Creek again changed its name, this time to Keyser, after William Keyser, the first vice president of the B & O Railroad.

No. 7607, a 2-8-8-4 Class EM-1 articulated locomotive, was photographed on July 7, 1948 at Cumberland, Maryland. Opposition to the 1864–1868 negotiations to build the line between Connellsville and Cumberland were finally overcome and construction began in 1868. The line was completed and opened in June 1871.

No. 7615, a 2-8-8-4 Class EM-1 articulated locomotive, was photographed on October 7, 1948 at Grafton, West Virginia. Early on, Grafton was recognized as an important junction point on the B & O Railroad. Trains from Wheeling, Columbus, Parkersburg, and Cincinnati intersected at Grafton.

In 1945, the Baldwin Locomotive Works built No. 7621, a 2-8-8-4 Class EM-1 articulated locomotive. It was photographed on May 12, 1947 at Grafton, West Virginia. The Cincinnati *Enquirer* wrote of the Grafton Hotel in 1857, "In point of convenience, beauty and all the et cetera that go to make up a hotel par excellence, the one at Grafton is not to be surpassed upon any line of railroad in this or any other country. The edifice is a grand structure composed of the Gothic and Corinthian style of architecture."

The Baldwin Locomotive Works built No. 7623, a 2-8-8-4 Class EM-1 articulated locomotive, as Baldwin No. 71505 in June 1945. It was photographed on May 11, 1949 at Keyser, West Virginia. New Creek was renamed Keyser to induce the B & O Railroad to move its facilities there.

No. 7624, a 2-8-8-4 Class EM-1 articulated locomotive, was built by the Baldwin Locomotive Works in 1945 and photographed on May 12, 1947, at Grafton, West Virginia. The 1911 Grafton B & O depot and the adjacent Willard Hotel were awarded $1 million in Federal Transportation grants for purchase and restoration.

No. 7625, a 2-8-8-4 Class EM-1 articulated locomotive, was photographed on June 4, 1949 at Cumberland, Maryland. The Class EM-1s were among the smallest Yellowstone locomotives in use. They were favored particularly because of the tight curves and clearances on the B & O lines.

2-10-0s

Alco Brooks built No. 6500, a 2-10-0 Class Y decapod-type locomotive, as No. 42720 in June 1907. Originally owned by the BR & P Railroad and operated as locomotive No. 501, it was photographed on April 23, 1947 at Keyser, West Virginia. This locomotive was scrapped in 1948. Note the engineer in the cab posing for the photographer.

2-10-2s

No. 535, a 2-10-2 locomotive, was photographed in New Castle, Pennsylvania on April 28, 1957. The Pittsburgh & Western Railroad had a total length of 205 miles including the line between Allegheny and New Castle, Pennsylvania. Construction began as a narrow gauge line between Etna and Zecienople in 1877 and opened in 1879.

On April 23, 1947, No. 6104, a 2-10-2 Class S-1 locomotive, was photographed at Cumberland, Maryland. President John W. Garrett oversaw the building of many hotels along the B & O routes including the Queen City Hotel in Cumberland in 1872. That same year a summer hotel at Deer Park, near the crest of the Allegheny Mountains, was built.

No. 6112, a 2-10-2 Class S-1 locomotive, was photographed at Keyser, West Virginia on April 23, 1947. This locomotive was built by Baldwin Locomotive Works as Baldwin No. 57338 in October 1923 and was retired from service in 1956.

Photographed on April 23, 1947, at Cumberland, Maryland, No. 6134 is a 2-10-2 Class S-1 locomotive. This locomotive, built by Baldwin Locomotive Works as Baldwin No. 57466 in November 1923, was originally numbered as No. 511 and was scrapped in 1958.

H. Newlon photographed No. 6142, a 2-10-2 Class S-1 locomotive, on July 18, 1936 at Glenwood, Pennsylvania. The B & O provided many volunteers for the 708th Railway Grand Division. They joined employees from 27 other railroads to form the newly organized Military Railway Service within the Army Transportation Corps.

No. 6199, a 2-10-2 Class S-1a locomotive, was photographed in July 1946 at Lima, Ohio. 2-10-2s were commonly used in pairs as helpers on the Pittsburg Division.

No. 6206, a 2-10-2 Class S-1a locomotive, was photographed on April 23, 1947 at Cumberland, Maryland. After completion of its work during the Normandy invasion, the 708th began restoration of rail service in France. The division also did yeoman's duty during the Battle of the Bulge in December 1944.

This is another shot of the previous locomotive, No. 6206, taken in July 1953 by H. Newlon at New Castle, Pennsylvania.

No. 6219, a 2-10-2 Class S-1a locomotive, was photographed on April 23, 1947, at Cumberland, Maryland. Ross Winan was an early builder for the B & O Railroad and, in 1845, built a locomotive named after the city of Cumberland. The *Cumberland* weighed in at 23-1/2 tons. The 0-8-0 locomotives of this style were called "mud diggers."

4-4-0s

Photographed at the Chicago Railroad Fair of 1948–1949 was No. 25, a 4-4-0 American-type locomotive. The Chicago Rail Fair was held during the summers of 1948 and 1949, and drew crowds as large as 50,000 per day with a total of 2.5 million visitors before the Railroad Fair was all over.

4-4-2s

No. 1424, a 4-4-2 Class A-3 locomotive, is shown in storage possibly waiting to be scrapped out. The date, location, and photographer are unknown.

No. 1, a 4-4-4 locomotive, *Lady Baltimore*, is being serviced at Bloomington, Illinois on June 10, 1935. It was photographed by R. J. Foster. *Lady Baltimore* was an experimental locomotive for the B & O. This was one of 3 or 4 "Jubilee"-type locomotives built. Note the European styling and enclosed cab.

4-6-0s

No. 117, a 4-6-0 ten-wheeler, *Thatcher Perkins*, is shown at the Chicago Railroad Fair in 1948–1849. Thatcher Perkins was a builder for the B & O Railroad and this was the first locomotive he turned out of the shops at Mount Clare in the spring of 1863. It sported bright red wheels and a similar stripe around the balloon stack while the tender was bright maroon. This locomotive was built for passenger service.

J. Boose photographed No. 178, a 4-6-0 Class B-58 ten-wheeler, on May 23, 1939, at Lima, Ohio. Lima was the home of Lima Locomotive Works. These locomotives were known as the Cadillacs of steam locomotives.

Photographed on September 5, 1938, while in storage at Parkersburg, West Virginia, is No. 1358, a 4-6-0 Class B-8 ten-wheeler. Note the sign above the second drive wheel, "DON'T TAKE ANY PARTS OFF THIS ENGINE." Perhaps this locomotive was going to see service in the future.

No. 2007, a 4-6-0 Class B-18d ten–wheeler, was photographed on March 23, 1947, at Zanesville, Ohio. Charles Goddard Dillon, was from one of the pioneer families in Zanesville. In 1849, he joined the Corps of Engineers surveying for the Central Ohio Railroad, which later became a part of the B & O Railroad.

Photographed on the turntable at Parkersburg, West Virginia, on April 25, 1947, is No. 2007, a 4-6-0 Class B-18d ten-wheeler. It was built by the Rhode Island Locomotive Works in 1901. In 1966, you could ride from Huntington, West Virginia, to Parkersburg in a bay window caboose; the ticket cost $6. Note the engineer in the cab.

No. 2020, a 4-6-0 Class B-18c ten-wheeler, was built by the Rhode Island Locomotive Works in 1901 and photographed on August 29, 1945, at Clifton, Staten Island, New York. On July 1, 1971, the City of New York purchased the Staten Island subsidiary of the B & O. The Staten Island Rapid Transit Operating Authority was created to operate the New York City Transit rail service on Staten Island.

No. 2022, a 4-6-0 Class B-18d ten-wheeler, was photographed on February 12, 1939, in Glenwood, Pennsylvania, by C. Dengler. This locomotive was built by Rhode Island Locomotive Works as a Class B-18 locomotive in August 1901. Sometime between 1908-1914 it was converted to a Class B-18d and was scrapped in September 1947. In 1932, tonnage on American railroads dropped to less than 50 percent of what it had been before the stock market crash in 1929. Tonnage again increased with the exception of 1938–1939 when a recession held back previous year's increases.

4-6-2s

Photographed at Cincinnati, Ohio, is No. 5038, a 4-6-2 Class P-1b Pacific-type locomotive. The Cincinnati, Hamilton, and Dayton Railroad was chartered on March 2, 1846. The railroad operated its first train on September 19, 1850; it eventually became a part of the B & O Railroad.

On July 7, 1940, R. J. Foster photographed No. 5039, a 4-6-2 Class P-1c Pacific-type locomotive in Cincinnati, Ohio. Cincinnati represented business for the B & O. Its residents had paid out $750,000 in subscriptions to pay for stock in the Cincinnati and Hamilton Railroad.

No. 5077 is a 4-6-2 Pacific-type locomotive. This locomotive was built in September 1917 as a Class P-1aa locomotive by the Baldwin Locomotive Works as Baldwin No. 36947. Originally numbered as B & O No. 2126, it was renumbered to the number shown in the photo in 1918 and was scrapped in April 1954.

No. 5078, a 4-6-2 Class P-1aa Pacific-type locomotive, was photographed on July 27, 1947 at Pittsburgh, Pennsylvania. The purchases of the Pittsburgh & Western, Pittsburgh, Cleveland & Toledo, and the Akron & Chicago Junction Railroads in 1891 gave the B & O a shorter route to Chicago.

H. Newlon photographed No. 5080, a 4-6-2 Class P-1-ba Pacific-type locomotive, on August 1, 1939, at Glenwood, Pennsylvania. This Class P-1 locomotive was built by Baldwin Locomotive Works as Baldwin No. 35903 in January 1911. It was rebuilt in 1943 as a Class T-3b locomotive.

No. 5124, a 4-6-2 Pacific-type locomotive, was photographed in March 1947 in Washington, D. C. The B & O line to Washington opened on August 24, 1835. The first locomotive used was the *Atlantic*. Four additional "grasshopper"-type locomotives, the *George Washington*, *John Adams*, *Thomas Jefferson*, and *James Madison* were used to transport 1,000 people from Baltimore to Washington on the first trip.

No. 5146 is a 4-6-2 Pacific-type locomotive. This locomotive was built by Alco Brooks Locomotive Works as Alco No. 53779 as a Class WW locomotive for BR & P railroad in July 1913. While on the BR & P the locomotive carried number 606. The B & O renumbered this locomotive to the number shown in the photo. Under B & O ownership this locomotive was listed as a Class P-17a and was sold for scrap in April 1952.

Photographed in 1938 is No. 5147, a 4-6-2 Class P-17-a Pacific-type locomotive, in Glenwood, Pennsylvania. This locomotive was built as BR & P railroad No. 607 as a Class WW by Alco Brooks as Alco No. 54707 in June 1914 and was sold for scrap in April 1952.

No. 5202 is a 4-6-2 Pacific-type locomotive. This locomotive was built as a Class P-5 locomotive by Baldwin Locomotive Works as Baldwin No. 51827 in June 1919. It was converted to a Class P-5a and scrapped in June 1954.

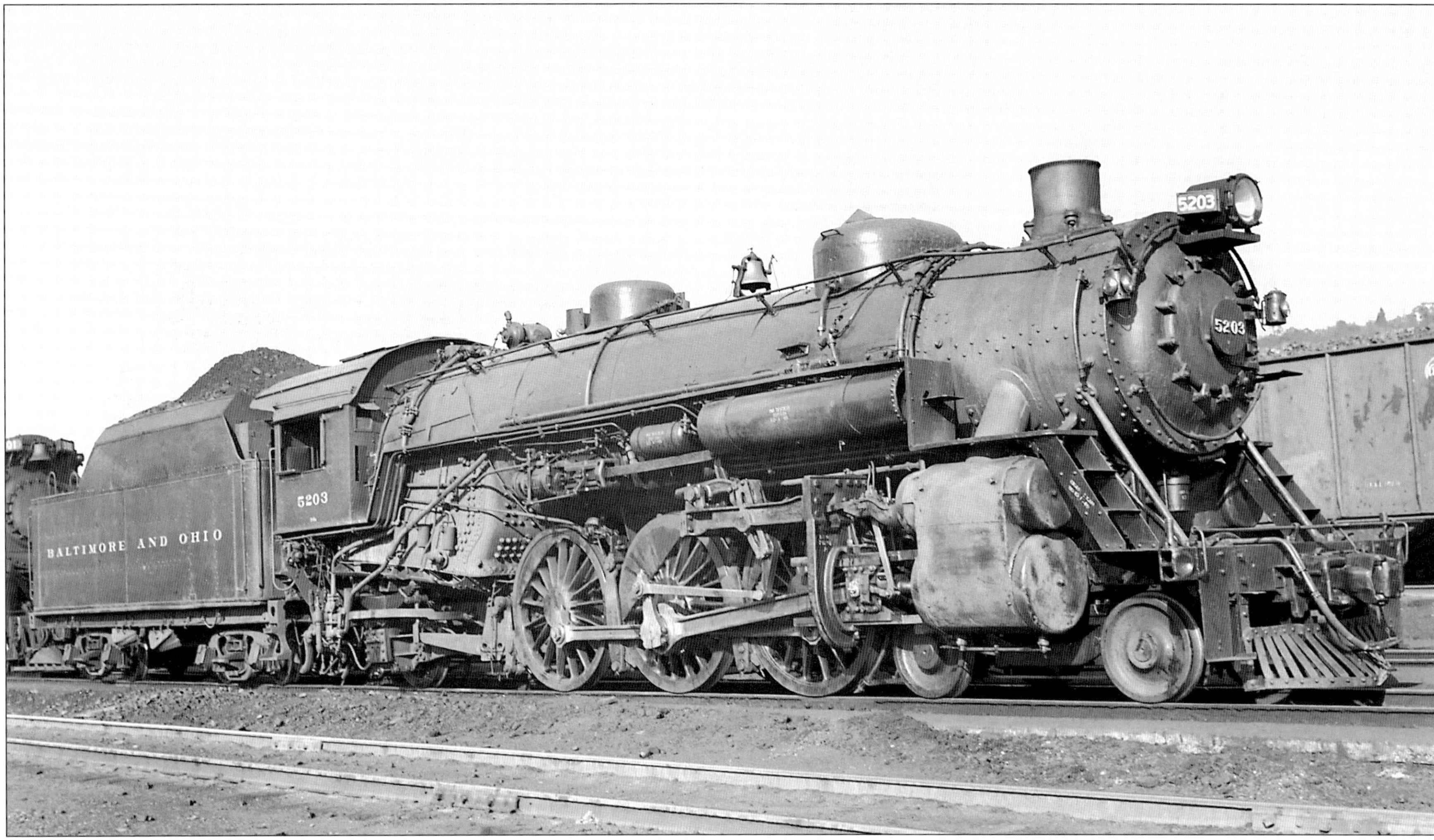

No. 5203, a 4-6-2 Pacific-type locomotive, was photographed on October 6, 1949, in Cincinnati, Ohio. An estimated 20,000 people attended the arrival of the first B & O train in Cincinnati. Early in the morning of that day the Cincinnati Fire Department and local militia held a large parade in commemoration.

No. 5214, a 4-6-2 Class P-5a Pacific-type locomotive, was photographed on April 27, 1947, at the K & IT yard in Louisville, Kentucky. The Baldwin Locomotive Works built this locomotive in 1919. John Jordan Upchurch, an employee of Baldwin Locomotive Works, was the founder of the Ancient Order of United Workmen, the first fraternal organization in America to offer insurance benefits to its members.

No. 5222 is a 4-6-2 Class P-5 Pacific-type locomotive. This Class P-5 locomotive was built by Alco Brooks as Alco No. 61278 in September 1919. It was converted to a Class P-5a in June 1929 and retired in October 1954.

Photographed on October 18, 1941, at Baltimore, Maryland, is No. 5226, a 4-6-2 Pacific-type locomotive. This Class P-5 locomotive was built by Alco Brooks as Alco No. 61282 in September 1916. It was converted to a Class P-5a locomotive in December 1928 and sold for scrap in July 1954.

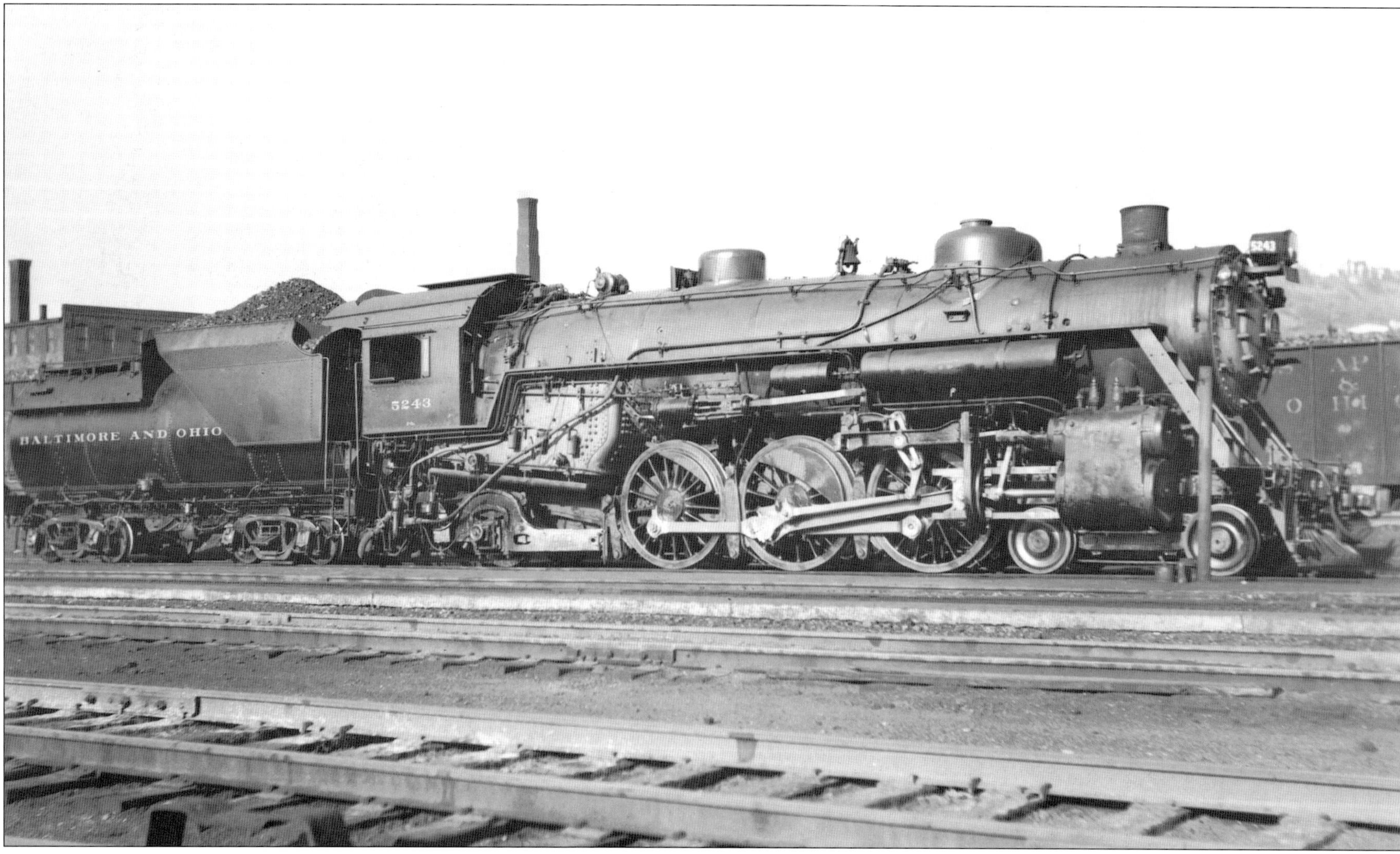

No. 5243, a 4-6-2 Class P-6 a Pacific-type locomotive, was photographed on April 28, 1947 in Cincinnati, Ohio. The Burnet House Hotel in Cincinnati was the site of a number of events marking the arrival of the first B & O train into the city. At that time, the Burnet House was known as one of the finest hotels in the world.

Shown in Cincinnati, Ohio, is No. 5302, a 4-6-2 Class P-7 Pacific-type locomotive. Four locomotives numbered 5301 to 5304 were rebuilt for the Presidential Series to be used on the new high-speed line linking Washington and Cincinnati. These locomotives used streamlined shrouds designed by Olive Dennis. The passenger cars used on these trains were painted royal blue, light gray, and black, with yellow striping and lettering.

No. 5305, a 4-6-2 Pacific-type locomotive, was photographed on November 3, 1953 in Communipaw, New Jersey. The B & O operated its passenger trains into Jersey City using track owned by the Reading Railroad and the Jersey Central Railroad.

The Baldwin Locomotive Works built locomotive No. 5311, a 4-6-2 Class P-7 Pacific-type, as Baldwin No. 59933 in April 1927. It was photographed on September 3, 1947 in Washington D. C. It is the ex-*President Fillmore*.

No. 5314 is a 4-6-2 Class P-7 Pacific-type locomotive. This Class P-7 locomotive was built by Baldwin Locomotive Works as Baldwin No. 59936 in April 1927. In 1949 it was converted to a Class P-7c with booster and retired in 1957. Locomotives numbered 5300-5319 were the B & O's presidential series of locomotives. This particular locomotive is the President Lincoln.

Photographed on November 26, 1949, in Philadelphia, Pennsylvania, is No. 5315, a 4-6-2 Class P-7e locomotive. This Class P-7 locomotive was built by the Baldwin Locomotive Works as Baldwin No. 59937 and rebuilt in 1949 as a Class P-7e and retired in 1956. This was the President Johnson locomotive.

4-6-4s

Robert J. Foster photographed No. 5320, a 4-6-4 Class P-7 Pacific-type locomotive, on May 31, 1954. Notice the passenger cars in the background. In August 1830, Richard Imlay of Philadelphia showed off his new "Improved Passenger Car." The *American* newspaper, wrote, "The carriage which is named the Ohio, is very handsomely finished and will no doubt, be a great favorite with the visitors to the railroad."

No. 5350, a 4-8-2 Class V-3 Hudson-type locomotive, was photographed on August 1, 1936 at New Castle Junction, Pennsylvania. This locomotive was built in the B & O Mount Clare Shops in June 1935. The Mount Clare Shops did the conversion work on the "improved Royal Blue heavyweight passenger cars."

No. 5501, a 4-8-2 Class T mountain-type locomotive, was photographed in June 1947 at Keyser, West Virginia. This locomotive was originally a B & O 2-10-2 Class T locomotive, No. 6030, and rebuilt at the Mount Clare Shops in January 1926.

No. 5510, a 4-8-2 mountain-type locomotive, was photographed in Chicago in June 1933 while on display at the Century of Progress Exposition. Judging by the photograph, this display by the B & O Railroad drew a lot of attention with visitors to the Exposition.

On April 19, 1953, R. J. Foster photographed No. 5560, a 4-8-2 mountain-type locomotive, at Riverdale, Illinois. When the B & O line to Chicago was completed, it was the most direct route from Washington to Chicago at 784 miles. This class T-3 locomotive was built in the B & O Mt. Clare shops in 1943. It was rebuilt and given the boiler from locomotive #5603 and was dropped from the roster in 1956.

No. 5562, a 4-8-2 Class 5 mountain-type locomotive, was photographed on August 24, 1949 at Willard, Ohio. In two years, 1910 and 1911, B & O President Daniel Willard purchased 434 locomotives to upgrade the power of the railroad. This purchase included 150 2-8-2, mikado-type locomotives from the Baldwin Locomotive Works.

Charles T. Felstead photographed No. 5574, a 4-8-2 mountain-type locomotive, on October 19, 1947, in Lima, Ohio. After the Lima Locomotive Works discontinued building steam locomotives, the company name changed to Lima-Hamilton Corporation and they began building diesel locomotives. The B & O purchased ten 1,000-horsepower switchers from Lima-Hamilton in the early 1950s.

The Baldwin Locomotive Works built No. 5653, a 4-8-2 Class T-4 mountain-type locomotive, as Baldwin No. 61825 in March 1935. It was photographed in July 1948 at Willard, Ohio. The locomotive originally belonged to the Boston & Maine Railroad and operated as No. 4103, a Class R-1a locomotive.

Converting to Diesel

No. 6038 is a Gas Electric unit and was photographed on January 1, 1952 at Zanesville, Ohio. Units such as this were used for self-contained/self-powered individual passenger service where ridership didn't call for a multiple car train.

Electro Motive Company's Box Cab, No. 50, is the *Abraham Lincoln* passenger train. Purchased in 1935, this unit was the first self-contained diesel passenger locomotive in the United States. In 1937, this locomotive was moved to the B & O's Alton Railroad and worked in lightweight passenger service between Chicago and St. Louis until the mid-1940s, finally retiring in 1958. Today the *Abraham Lincoln* is in the St. Louis Transportation Museum.

No. 51, an EMC EA passenger train, is preparing to leave St. Louis Union Station. Note the engineer standing along side the locomotive. For many years, St. Louis's Union Station fell into disrepair and was abandoned. It has been restored and is a vibrant business mall with many shops. The Amtrak stop for St. Louis is located about three blocks from Union Station.

AMERICAN CULTURE

Coca-Cola: A History in Photographs 1930-1969 ISBN 1-882256-46-8
Coca-Cola: Its Vehicles in Photographs 1930-1969 ISBN 1-882256-47-6
Phillips 66 1945-1954 Photo Archive ISBN 1-882256-42-5

AUTOMOTIVE

AMX Photo Archive: From Concept to Reality ISBN 1-58388-062-3
Auburn Automobiles 1900-1936 Photo Archive ISBN 1-58388-093-3
Camaro 1967-2000 Photo Archive ISBN 1-58388-032-1
Checker Cab Co. Photo History ISBN 1-58388-100-X
Chevrolet Station Wagons 1946-1966 Photo Archive ISBN 1-58388-069-0
Classic American Limousines 1955-2000 Photo Archive ISBN 1-58388-041-0
Corvair by Chevrolet Experimental & Production Cars 1957-1969, Ludvigsen Library Series ISBN 1-58388-058-5
Corvette The Exotic Experimental Cars, Ludvigsen Library Series ISBN 1-58388-017-8
Corvette Prototypes & Show Cars Photo Album ISBN 1-882256-77-8
Early Ford V-8s 1932-1942 Photo Album ISBN 1-882256-97-2
Ferrari- The Factory Maranello's Secrets 1950-1975, Ludvigsen Library Series ISBN 1-58388-085-2
Ford Postwar Flatheads 1946-1953 Photo Archive ISBN 1-58388-080-1
Ford Station Wagons 1929-1991 Photo History ISBN 1-58388-103-4
Imperial 1955-1963 Photo Archive ISBN 1-882256-22-0
Imperial 1964-1968 Photo Archive ISBN 1-882256-23-9
Javelin Photo Archive: From Concept to Reality ISBN 1-58388-071-2
Lincoln Motor Cars 1920-1942 Photo Archive ISBN 1-882256-57-3
Lincoln Motor Cars 1946-1960 Photo Archive ISBN 1-882256-58-1
Nash 1936-1957 Photo Archive ISBN 1-58388-086-0
Packard Motor Cars 1935-1942 Photo Archive ISBN 1-882256-44-1
Packard Motor Cars 1946-1958 Photo Archive ISBN 1-882256-45-X
Pontiac Dream Cars, Show Cars & Prototypes 1928-1998 Photo Album ISBN 1-882256-93-X
Pontiac Firebird Trans-Am 1969-1999 Photo Album ISBN 1-882256-95-6
Pontiac Firebird 1967-2000 Photo History ISBN 1-58388-028-3
Rambler 1950-1969 Photo Archive ISBN 1-58388-078-X
Stretch Limousines 1928-2001 Photo Archive ISBN 1-58388-070-4
Studebaker 1933-1942 Photo Archive ISBN 1-882256-24-7
Studebaker Hawk 1956-1964 Photo Archive ISBN 1-58388-094-1
Studebaker Lark 1959-1966 Photo Archive ISBN 1-58388-107-7
Ultimate Corvette Trivia Challenge ISBN 1-58388-035-6

BUSES

Buses of ACF Photo Archive ISBN 1-58388-101-8
Buses of Motor Coach Industries 1932-2000 Photo Archive ISBN 1-58388-039-9
Fageol & Twin Coach Buses 1922-1956 Photo Archive ISBN 1-58388-075-5
Flxible Intercity Buses 1924-1970 Photo Archive ISBN 1-58388-108-5
Flxible Transit Buses 1953-1995 Photo Archive ISBN 1-58388-053-4
GM Intercity Coaches 1944-1980 Photo Archive ISBN 1-58388-099-2
Greyhound Buses 1914-2000 Photo Archive ISBN 1-58388-027-5
*Mack® Buses 1900-1960 Photo Archive** ISBN 1-58388-020-8
Prevost Buses 1924-2002 Photo Archive ISBN 1-58388-083-6
Trailways Buses 1936-2001 Photo Archive ISBN 1-58388-029-1
Trolley Buses 1913-2001 Photo Archive ISBN 1-58388-057-7
Yellow Coach Buses 1923-1943 Photo Archive ISBN 1-58388-054-2

EMERGENCY VEHICLES

The American Ambulance 1900-2002: An Illustrated History ISBN 1-58388-081-X
American Funeral Vehicles 1883-2003 Illustrated History ISBN 1-58388-104-2
American LaFrance 700 Series 1945-1952 Photo Archive ISBN 1-882256-90-5
American LaFrance 700 Series 1945-1952 Photo Archive Volume 2 ISBN 1-58388-025-9
American LaFrance 700 & 800 Series 1953-1958 Photo Archive ISBN 1-882256-91-3
American LaFrance 900 Series 1958-1964 Photo Archive ISBN 1-58388-002-X
Classic Seagrave 1935-1951 Photo Archive ISBN 1-58388-034-8
Crown Firecoach 1951-1985 Photo Archive ISBN 1-58388-047-X
Fire Chief Cars 1900-1997 Photo Album ISBN 1-882256-87-5
Hahn Fire Apparatus 1923-1990 Photo Archive ISBN 1-58388-077-1
Heavy Rescue Trucks 1931-2000 Photo Gallery ISBN 1-58388-045-3
Imperial Fire Apparatus 1969-1976 Photo Archive ISBN 1-58388-091-7
Industrial and Private Fire Apparatus 1925-2001 Photo Archive ISBN 1-58388-049-6
Los Angeles City Fire Apparatus 1953-1999 Photo Archive ISBN 1-58388-012-7
*Mack Model C Fire Trucks 1957-1967 Photo Archive** ISBN 1-58388-014-3
*Mack Model L Fire Trucks 1940-1954 Photo Archive** ISBN 1-882256-86-7
Maxim Fire Apparatus 1914-1989 Photo Archive ISBN 1-58388-050-X
Navy & Marine Corps Fire Apparatus 1836 -2000 Photo Gallery ISBN 1-58388-031-3
Pierre Thibault Ltd. Fire Apparatus 1918-1990 Photo Archive ISBN 1-58388-074-7
Pirsch Fire Apparatus 1890-1991 Photo Archive ISBN 1-58388-082-8
Police Cars: Restoring, Collecting & Showing America's Finest Sedans ISBN 1-58388-046-1
Saulsbury Fire Rescue Apparatus 1956-2003 Photo Archive ISBN 1-58388-106-9
Seagrave 70th Anniversary Series Photo Archive ISBN 1-58388-001-1
TASC Fire Apparatus 1946-1985 Photo Archive ISBN 1-58388-065-8
Volunteer & Rural Fire Apparatus Photo Gallery ISBN 1-58388-005-4
W.S. Darley & Co. Fire Apparatus 1908-2000 Photo Archive ISBN 1-58388-061-5
Ward LaFrance Fire Trucks 1918-1978 Photo Archive ISBN 1-58388-013-5
Wildland Fire Apparatus 1940-2001 Photo Gallery ISBN 1-58388-056-9
Young Fire Equipment 1932-1991 Photo Archive ISBN 1-58388-015-1

RACING

Chaparral Can-Am Racing Cars from Texas, Ludvigsen Library Series ISBN 1-58388-066-6
Cunningham Sports Cars, Ludvigsen Library Series ISBN 1-58388-109-3
Drag Racing Funny Cars of the 1960s Photo Archive ISBN 1-58388-097-6
Drag Racing Funny Cars of the 1970s Photo Archive ISBN 1-58388-068-2
El Mirage Impressions: Dry Lakes Land Speed Racing ISBN 1-58388-059-3
GT40 Photo Archive ISBN 1-882256-64-6
Indy Cars of the 1950s, Ludvigsen Library Series ISBN 1-58388-018-6
Indy Cars of the 1960s, Ludvigsen Library Series ISBN 1-58388-052-6
Indy Cars of the 1970s, Ludvigsen Library Series ISBN 1-58388-098-4
Indianapolis Racing Cars of Frank Kurtis 1941-1963 Photo Archive ISBN 1-58388-026-7
Juan Manuel Fangio World Champion Driver Series Photo Album ISBN 1-58388-008-9

More Great Titles From Iconografix

All Iconografix books are available from direct mail specialty book dealers and bookstores worldwide, or can be ordered from the publisher. For book trade and distribution information or to add your name to our mailing list and receive a **FREE CATALOG** contact:

Iconografix,
PO Box 446, Dept BK
Hudson, WI, 54016

Telephone: (715) 381-9755,
(800) 289-3504 (USA),
Fax: (715) 381-9756

Lost Race Tracks Treasures of Automobile Racing ISBN 1-58388-084-4
Mario Andretti World Champion Driver Series Photo Album ISBN 1-58388-009-7
Mercedes-Benz 300SL Racing Cars 1952-1953, Ludvigsen Library Series ISBN 1-58388-067-4
Novi V-8 Indy Cars 1941-1965, Ludvigsen Library Series ISBN 1-58388-037-2
Porsche Spyders Type 550 1953-1956, Ludvigsen Library Series ISBN 1-58388-092-5
Sebring 12-Hour Race 1970 Photo Archive ISBN 1-882256-20-4
Vanderbilt Cup Race 1936 & 1937 Photo Archive ISBN 1-882256-66-2

RAILWAYS

Chicago, St. Paul, Minneapolis & Omaha Railway 1880-1940 Photo Archive ISBN 1-882256-67-0
Chicago & North Western Railway 1975-1995 Photo Archive ISBN 1-882256-76-X
Great Northern Railway 1945-1970 Volume 2 Photo Archive ISBN 1-882256-79-4
Great Northern Railway Ore Docks of Lake Superior Photo Archive ISBN 1-58388-073-9
Illinois Central Railroad 1854-1960 Photo Archive ISBN 1-58388-063-1
Milwaukee Road 1850-1960 Photo Archive ISBN 1-882256-61-1
Milwaukee Road Depots 1856-1954 Photo Archive ISBN 1-58388-040-2
Show Trains of the 20th Century ISBN 1-58388-030-5
Soo Line 1975-1992 Photo Archive ISBN 1-882256-68-9
Steam Locomotives of the B&O Railroad Photo Archive ISBN 1-58388-095-X
Streamliners to the Twin Cities Photo Archive 400, Twin Zephyrs & Hiawatha Trains ISBN 1-58388-096-8
Trains of the Twin Ports Photo Archive, Duluth-Superior in the 1950s ISBN 1-58388-003-8
Trains of the Circus 1872-1956 ISBN 1-58388-024-0
Trains of the Upper Midwest Photo Archive Steam & Diesel in the 1950s & 1960s ISBN 1-58388-036-4
Wisconsin Central Limited 1987-1996 Photo Archive ISBN 1-882256-75-1
Wisconsin Central Railway 1871-1909 Photo Archive ISBN 1-882256-78-6

RECREATIONAL VEHICLES

Ski-Doo Racing Sleds 1960-2003 Photo Archive ISBN 1-58388-105-0

TRUCKS

Autocar Trucks 1950-1987 Photo Archive ISBN 1-58388-072-0
Beverage Trucks 1910-1975 Photo Archive ISBN 1-882256-60-3
*Brockway Trucks 1948-1961 Photo Archive** ISBN 1-882256-55-7
Chevrolet El Camino Photo History Incl. GMC Sprint & Caballero ISBN 1-58388-044-5
Circus and Carnival Trucks 1923-2000 Photo Archive ISBN 1-58388-048-8
Dodge B-Series Trucks Restorer's & Collector's Reference Guide and History ISBN 1-58388-087-9
Dodge Pickups 1939-1978 Photo Album ISBN 1-882256-82-4
Dodge Power Wagons 1940-1980 Photo Archive ISBN 1-882256-89-1
Dodge Power Wagon Photo History ISBN 1-58388-019-4
Dodge Ram Trucks 1994-2001 Photo History ISBN 1-58388-051-8
Dodge Trucks 1929-1947 Photo Archive ISBN 1-882256-36-0
Dodge Trucks 1948-1960 Photo Archive ISBN 1-882256-37-9
Ford 4x4s 1935-1990 Photo History ISBN 1-58388-079-8
Ford Heavy-Duty Trucks 1948-1998 Photo History ISBN 1-58388-043-7
Freightliner Trucks 1937-1981 Photo Archive ISBN 1-58388-090-9
Jeep 1941-2000 Photo Archive ISBN 1-58388-021-6
Jeep Prototypes & Concept Vehicles Photo Archive ISBN 1-58388-033-X
*Mack Model AB Photo Archive** ISBN 1-882256-18-2
*Mack AP Super-Duty Trucks 1926-1938 Photo Archive** ISBN 1-882256-54-9
*Mack Model B 1953-1966 Volume 2 Photo Archive** ISBN 1-882256-34-4
*Mack EB-EC-ED-EE-EF-EG-DE 1936-1951 Photo Archive** ISBN 1-882256-29-8
*Mack EH-EJ-EM-EQ-ER-ES 1936-1950 Photo Archive** ISBN 1-882256-39-5
*Mack FC-FCSW-NW 1936-1947 Photo Archive** ISBN 1-882256-28-X
*Mack FG-FH-FJ-FK-FN-FP-FT-FW 1937-1950 Photo Archive** ISBN 1-882256-35-2
*Mack LF-LH-LJ-LM-LT 1940-1956 Photo Archive** ISBN 1-882256-38-7
*Mack Trucks Photo Gallery** ISBN 1-882256-88-3
New Car Carriers 1910-1998 Photo Album ISBN 1-882256-98-0
Plymouth Commercial Vehicles Photo Archive ISBN 1-58388-004-6
Refuse Trucks Photo Archive ISBN 1-58388-042-9
RVs & Campers 1900-2000: An Illustrated History ISBN 1-58388-064-X
Studebaker Trucks 1927-1940 Photo Archive ISBN 1-882256-40-9
White Trucks 1900-1937 Photo Archive ISBN 1-882256-80-8

TRACTORS & CONSTRUCTION EQUIPMENT

Case Tractors 1912-1959 Photo Archive ISBN 1-882256-32-8
Caterpillar Photo Gallery ISBN 1-882256-70-0
Caterpillar Pocket Guide The Track-Type Tractors 1925-1957 ISBN 1-58388-022-4
Caterpillar D-2 & R-2 Photo Archive ISBN 1-882256-99-9
Caterpillar D-8 1933-1974 Photo Archive Incl. Diesel 75 & RD-8 ISBN 1-882256-96-4
Caterpillar Military Tractors Volume 1 Photo Archive ISBN 1-882256-16-6
Caterpillar Military Tractors Volume 2 Photo Archive ISBN 1-882256-17-4
Caterpillar Sixty Photo Archive ISBN 1-882256-05-0
Caterpillar Ten Photo Archive Incl. 7c Fifteen & High Fifteen ISBN 1-58388-011-9
Caterpillar Thirty Photo Archive 2ND Ed. Incl. Best Thirty, 6G Thirty & R-4 ISBN 1-58388-006-2
Circus & Carnival Tractors 1930-2001 Photo Archive ISBN 1-58388-076-3
Cletrac and Oliver Crawlers Photo Archive ISBN 1-882256-43-3
Classic American Steamrollers 1871-1935 Photo Archive ISBN 1-58388-038-0
Farmall Cub Photo Archive ISBN 1-882256-71-9
Farmall F-Series Photo Archive ISBN 1-882256-02-6
Farmall Model H Photo Archive ISBN 1-882256-03-4
Farmall Model M Photo Archive ISBN 1-882256-15-8
Farmall Regular Photo Archive ISBN 1-882256-14-X
Farmall Super Series Photo Archive ISBN 1-882256-49-2
Fordson 1917-1928 Photo Archive ISBN 1-882256-33-6
Hart-Parr Photo Archive ISBN 1-882256-08-5
Holt Tractors Photo Archive ISBN 1-882256-10-7
International TracTracTor Photo Archive ISBN 1-882256-48-4
John Deere Model A Photo Archive ISBN 1-882256-12-3
John Deere Model D Photo Archive ISBN 1-882256-00-X
Marion Construction Machinery 1884-1975 Photo Archive ISBN 1-58388-060-7
Marion Mining & Dredging Machines Photo Archive ISBN 1-58388-088-7
Oliver Tractors Photo Archive ISBN 1-882256-09-3
Russell Graders Photo Archive ISBN 1-882256-11-5
Twin City Tractor Photo Archive ISBN 1-882256-06-9

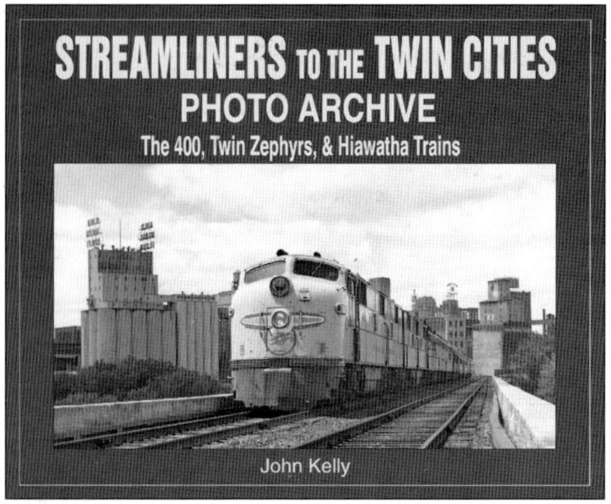

More great books from

Iconografix

Streamliners to the Twin Cities Photo Archive
ISBN 1-58388-096-8

Great Northern Railway Ore Docks of Lake Superior Photo Archive
ISBN 1-58388-073-9

Illinois Central Railroad 1854-1960 Photo Archive
ISBN 1-58388-063-1

Milwaukee Road 1850-1960 Photo Archive
ISBN 1-882256-61-1

Milwaukee Road Depots 1856-1954 Photo Archive
ISBN 1-58388-040-2

Trains of the Upper Midwest Photo Archive
ISBN 1-58388-036-4

Wisconsin Central Limited 1987-1996 Photo Archive
ISBN 1-882256-75-1

Iconografix, Inc.
P.O. Box 446, Dept BK,
Hudson, WI 54016
For a **free catalog** call: 1-800-289-3504

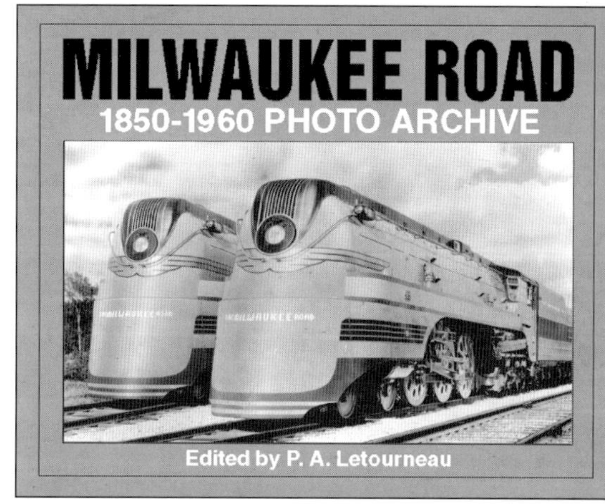

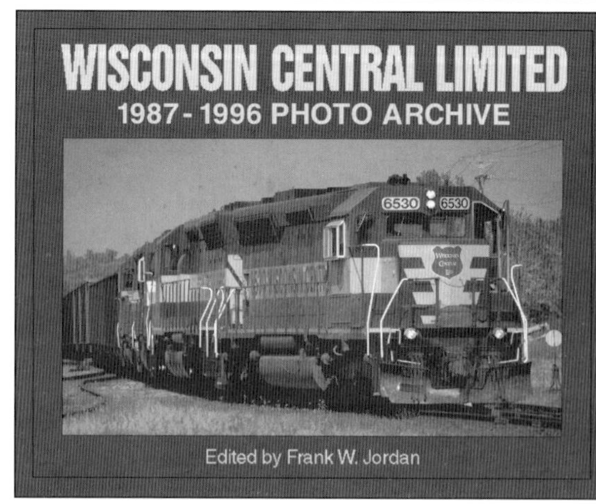